SWALLOWING  GEOGRAPHY

Deborah Levy was born in 1959, studied theatre at Dartington College of Arts, and now lives in London. Her plays include *Pax* and *Heresies* for the Royal Shakespeare Company. Her publications include poetry, *An Amorous Discourse in the Suburbs of Hell*, short stories, *Ophelia and the Great Idea* and a novel, *Beautiful Mutants* about which the *Observer* commented: 'it throbs its way into the imagination like the unguided missile it decries'.

'*Swallowing Geography* continues Levy's exploration of these "splintered times"... Levy has rejected conventional narrative structure in favour of a poetic sequence... It shifts effortlessly between different voices and scenes, between the inner and the outer life. This rhythmic flow of images and events is anchored by the author's keen eye for detail, and the patience and conviction with which she constructs the whole. There is a continual change of pace and tone, as concise description and terse conversation give way to impassioned monologues and philosophical discourse... This is a world where boundaries blur, where the order of everyday life is constantly invaded by the inner chaos of the subconscious or the wider chaos of international events. The book is studded with references to Yugoslavia, Chernobyl and the Middle East. Palm trees at a tacky beach resort bring to mind war in the desert; a radio newscast lists the civilian dead, as JK watches the sun set over the sea. These events are contemporary points of reference, signposts and symbols that are part of the "geography" Levy is attempting to describe... Deborah Levy has achieved far more here than a simple study of contemporary disintegration and malaise... some truly original writing reflects, in its careful selection and placing of parts, the fact that, despite the illusory nature of geography, we persist in looking for new ways to draw the map'

Times Literary Supplement

'It is rare to find an author whose work one awaits with greedy anticipation, and whose writing fills one with excitement and gratitude. Deborah Levy is such an author. She writes with an originality and poetic intensity which takes the breath away. Her subject matter is always apocalyptic – a knife-edge plunge into the millennial chasm of the late twentieth century – and her use of language is always rich, sensual and evocative. She is a bold experimenter, so neither a linear narrative nor an everyday storyline should be expected. Her writing has a surreal beauty, a wildness, which is intoxicating... She offers herself as an eerie but reliable helmswoman to steer you through the stormy waters of our century's final, fear-filled days. It takes a leap of imagination and faith to enter her world – or, rather, our world, through her eyes – but it is a risk worth taking. Go on, jump'

Tribune

Deborah Levy

SWALLOWING GEOGRAPHY

VINTAGE

VINTAGE
20 Vauxhall Bridge Road, London SW1V 2SA

London Melbourne Sydney Auckland Johannesburg
and agencies throughout the world

First published by Jonathan Cape Ltd, 1993

Vintage edition 1993

2 4 6 8 10 9 7 5 3 1

© Deborah Levy 1993

Printed and bound in Great Britain by
Cox & Wyman, Reading, Berkshire

ISBN 0 09 920001 5

CONTENTS

And I say to any man or woman, Let your soul stand
cool and composed before a million universes.

Do I contradict myself?
Very well then I contradict myself
I am large, I contain multitudes

Walt Whitman, *Song of Myself*

I

THE TADPOLE FIELDS

'When you feel fear, does it have detail or is it just a force?' The gold filling in Gregory's front tooth shines into J.K.'s eye. 'I can't hear you.'

They are sitting in a bar surrounded by mirrors etched with the Eiffel Tower at Roissy airport, Paris. Brand names like Segrafredo, Perrier, Dior, Kronenbourg 1664, Chanel, spin like planets above them. The large blonde Californian waitress slams two cocktails down on the table.

'They're killers,' she says.

'It is as if Paris is muffled. I hear it at low volume.'

J.K. says nothing because these days Gregory's voice is very quiet, as if frightened to let it out of his big body. She catches odd words like nausea, chestpain, The Baltic States, mother, father, aspirin, and sometimes she catches his eye.

'Look at the inscription in this book.'

She moves closer to the secret in his body that tames his voice. It is an old volume of short stories printed in 1941 on thin transparent paper.

'Leave this book at the Post Office when you have read it, so that men and women in the services may enjoy it too.'

The day before, they walked to Pigalle in silence, arm in arm, stopping to watch transvestite whores lean against cars and walls, put on lipstick, smoke cigarettes, call out to men passing by, their steamy drugged gaze settling on this man and that man and then somewhere else.

'What kind of cultural virus taught those boys to stick their hips out like that, and pout and press their breasts across the other side of the road?' Gregory says.

'Do you fancy them?'

'They're gorgeous. I like that one over there with the long black plait . . . it comes down to his knees, can you see . . . in the hat with the red feather.'

COME AND TALK TO ME! It is an eerie staccato voice. The voice of cigarette advertisements, fierce sun, a two-bit bar with dead flies on the floor. They turn round to see they are standing outside a pinball arcade, CASINO spelt in coloured lightbulbs above the door. COME AND TALK TO ME! The yankee growling voice comes from a silver and chrome machine. On its screen a square muscled man jumps up and down in a computerised urban landscape of skyscrapers and highways. Hands in raincoat pocket, jaw jerking to one side, he drawls again, COME AND TALK TO ME!

Gregory nudges J.K.'s arm. 'Well listen to the man, let's take up his invitation.' He puts ten francs in the slot.

HOW YA DOING? says the man. TYPE SOMETHING INTO THE KEYBOARD AND I WILL RESPOND. The screen whirs as the urban cowboy crosses his arms and leans towards them.

'It's in English.'

'Well, tell him how you are.'

Gregory turns to face the man. He puts another ten francs into the machine and spreads out his fingers. Pink and blue bulbs flash above him.

Do your lips burn up when kissed right? Let me kiss 'em baby. Let me let me let me. I would like to fuck you. I would like to make you happy. How do you like to be touched? On the aeroplane over here, the air hostess demonstrated various ways of surviving an aircrash. She said we must blow on a whistle to draw attention to ourselves. Don't you think that is a little narcissistic? If everyone in the everyday of their lives who wanted to draw attention to themselves blew a whistle where would we be? What do you do to make people love you? I do cheap things to make people like me. I make them feel more important than they are and flatter them and when someone makes me a great cocktail I take a sip and shout DRAGONFLIES! In England I light my cigarettes with matches made in Yugoslavia. The picture on the box is of 'Scenic Cornwall' and shows a number of signposts on the edge of a cliff. One of them says THE FALKLANDS 8109 and the other says AUSTRALIA 170001. I

3

tell you this because when I was a boy I collected stamps. It was my way of naming placing and conquering the world. A stamp is a small picture. So I had lots of small pictures of the world. Madagascar, China, Mexico, Argentina, Egypt. A kind of virtual reality.

What's your name, my sweet? Is it Johnny or Sam or Brett? I'd like to go down on you and for you to talk to me about football and religion and hamburgers and beauty and death and what it feels like to come. Were you bullied at school? When you were a teenager did you spend hours in your bedroom changing your clothes? Did you save up to buy the boots and shirts other kids had? What kind of Darwinian programmed you? Do you want to change yourself in any way? Like speak in a deeper voice or have a different nose? Do you feel safe in this world? Or do you feel alone and scared? What kind of gadgets do you have in your home? Do they comfort you? Baby do you sometimes feel glum? Baby take care of yourself. Oh baby I'd like to stroke you and whisper things to you and make you not have fear.

Honey, I want to tell you about a train I took to Kiev with my bit of squeeze. We made love just as we got near Chernobyl and the loudspeakers in our carriage played a kind of lament to mark the tragedy of the nuclear accident. In some way it seemed to mark all tragedy ever. The cries of our lovemaking as we passed infected cattle, children with shaved heads playing by the railway tracks and the eerie stillness of deformed trees was the only sound, snow falling, he and I sweating in each other's arms and honey we were, at that moment, without fear. The high-rise blocks of flats we stayed in were called The Sleeping Region. I was brought up in a block like that in London. As a kid we lived on tins of beans and meatballs and hated to sleep because we were frightened. Darling, do you sleep sweet and easy and deep? Does someone sleep beside you? Breathing into the pillow next to you and you wake up first and feel them there and it's just so great that they're there and you know very soon they will wake too and you will move closer and kind of pull in the beginning of a new day together? In Kiev I opened tins of crab meat and caviar bought with hard currency and we slept easy. We slept easy and there was a famine outside. The circus played every night in Kiev – an old man sitting next to me made a joke about eating the cats and horses after the show. Are you happy with your life, my sweet? The man said, 'You can always tell a tourist, their eyes don't know where they're going. Here everyone knows where they're going.' Do you know where you're going, baby? Is it a good place? Something to write home about? Is home a good place? Or just somewhere to return to?

Are you pleased to open your eyes in the morning? What do you see? Do you like what you see? If you hate it do you feel you have any power to change it for something else? Oh my love, let me call you that – My Love – let us imagine what that means, you and I liplocked some place in the American South, perhaps where the Klan lynched our brothers? You and I in a motor on the highway making plans for the future. The radio is on and we hear the Soviet Union has come apart and then there are some ads for Pepsi and bagel chips, and back to a war in Yugoslavia, nationalisms and internationalisms, an election in Great Britain, refugees crossing mountains looking for a country to feed them, a jingle for vitamin capsules, and all the time we are hot for each other, through all this world news we just want to be in each other's pants, and we pull in for gas and I'm saying, No baby don't light a cigarette right now, wait till we pull out and anyhow we'll check into a motel soon. Hey Brett, I'm Imagining America! It's all from movies and magazines, I'm fumbling to make you America. I'm fumbling to make you and unmake you. Abe Lincoln on your dollar bills – IN GOD WE TRUST – pastrami and gas and tacos and beer bought with his image, he's the guy that keeps the wheels turning. I'm stuffing chocolate into your mouth and baby . . . you're so hard, so hard honey . . . you're all fired up and I'm talkin' dirty, I'm talking physical, I'm talking politics and dontcha just love it, got my fingers in your armpit and you're sweating bad. I want you too, baby, I want you too. Y'know that Springsteen song . . . oh baby I'd drive all night again jus' to buy you a pair of shoes? Well I would. I'd drive to hell and back jus' to make you love me. How do you love? Do you keep it quiet and put it all in your fingertips or do you say words? What are your lovewords baby? What if the United States came apart? Would God come apart too and the stone pillars of the Abe Lincoln memorial crumble and statues of George Washington be torn up from squares of green,

6

watered by sprinklers? Torn up by crane and bulldozer?

Now I am imagining Switzerland, Brett. I can see snow
and stripped pine floors and coffee shops and cream cakes
and blond people tinkling little silver spoons against their
cups. I see children in nursery schools that are heated, very
warm and very clean and their little snow boots lined up
against the wall and gloves sewn into their coat pockets. I
can't imagine you there, Brett. I'm trying to see a teacher
bent over your shoulder while you draw your mother and
father and the house you live in and giant flowers – but
I just can't vision you in Switzerland skiing and eating
chocolate. You'd probably shoot up in your chalet, lie
down in your shorts under the skylight, arms folded behind
your neck looking up at the stars dreaming of home and
bourbon and cookies and having a haircut. You see how
I'm making you up, same as Switzerland and America?
Does it feel like it fits you? Have you made me up too?
Am I some kind of English faggot crazy for boys, cruising
into my adult life in black leather under strobe and sonic
boom of city discos? There's such a lot to talk about baby,
just you and me, man to man. Did you hear about the man
who went to a psychologist and said, doctor I think I'm
a dog, and the doctor said, we'll soon sort that out, now
get on the couch. And the man said, but I'm not allowed.
Well I'm inviting you to be whatever you like sweetheart,
I'm listening to you, I'm listening to everything you want
to be and were not allowed. Brett, I'm saying make yourself up for me baby, have as many goes as you like, be the
man you always wanted to be, and I'll be the man that

lets you. Brett, life is long dontcha think? When you tot up the hours and days and months it's a lot of time. How much of that time have you felt precious? I want to make you feel precious, my treasure, my lovestuff. Have you ever driven across a city you don't know very well and you're alone? It's night and you're lost. Had too many beers in some bar where they look at you as if you're an extra-terrestrial immigrant and somewhere else, in another city, there's someone who loves you and you imagine them looking at you in this bar now, checking you out, what shoes you've put on today and what you're drinking and what kind of mood you're in? And you want to say to the people in this bar who think you're some kind of weirdo blown in to undo them – I am connected to the same things as you y'know – I have people who love me and I watch TV and I have a birthday and I brush my teeth and I'm not always like this, eating crap pizza alone and lost with this look in my eyes. And then you get into the car and none of the street signs make sense, and you just cry. Brett, have you done this? And you think of all the people you've jilted meanly and all the people who dumped you, and your pockets are full of old bills and tickets and you turn over all the secrets you carry inside you?

SOUNDS LIKE YA NEED SOME HELP! The handsome urban cowboy uncrosses his thick arms and takes out a gun. Suddenly he jumps onto a moving car, shoots, jumps off the car and thrashes a man across the head with his gun, runs, leaps over a motorcycle, crouches, shoots, climbs up a skyscraper, hangs on with one hand, shoots with the

8

other, kicks a man chasing him off the building, shoots him in mid-air, dives through a pane of glass, shoots two three four five six other men, runs onto the roof of the skyscraper, flags down a helicopter, gets into it and pours bullets into the heart of the city – a loop of shooting and dying and dying and shooting and shooting and dying and then the voice says ... COME AND TALK TO ME ...

'I can't hear you, Greg,' J.K. says in the airport lounge.

'I know.'

'Have you seen a doctor?'

'Yeah.'

They finish their cocktails in silence.

'Look.'

J.K. opens her mouth and shows him the bloody gap where a tooth should be.

Gregory stares at her. Black flecks float in his green eyes.

'What happened?'

'I got slugged.'

'By who?'

'My mother. Knocked a tooth out.'

'Lillian?'

Their flight number comes up on the screen and they walk to the gate for take off.

'Tell me about getting slugged.'

The aeroplane shudders and they put on their seat belts.

2

THE TERRIBLE RAGES OF
LILLIAN STRAUSS

Her mother comes towards her carrying a black suit-case. J.K. walks three paces, takes the bag from her mother's hands and says, when I walked towards you, something inside me walked too. Some beast that has taken up residence inside me, that inhabits the colony of my interior, that has decided to inhabit me against my will, walked with me towards you. It lifted up each of its four legs and walked the three paces with me. The petrol winds from the Texaco petrol station wake her up blowing through the open window, and instead of birds, the panicky whir of the carwash. Later, on her balcony, J.K. sees the camellia she planted last year has blossomed. One pink bud, opening and opening, just as she in sleep is opening and her mother sliding in behind her eyes.

Lillian Strauss is J.K.'s mother.

Her mother says, in Singapore every evening at six o'clock, a tray of drinks would be brought in by a servant. Gin and tonic, an ice bucket, lemon and bitters. We would drink until we went to bed, dropped on cotton pillows. Your uncle had a grog's blossom for a nose, purple-veined, a great bloom. J.K. thinks of her camellia blossoming on a concrete balcony in a poor part of London, opening and opening.

Her mother, in another poor part of London, walks the streets in one of her rages. She is mouthing words to lamp-posts and parked cars; she is weeping and she is drunk. Her husbands have left her, her children have left her, her beauty has left her, she is filling the holes of absence with gin-scented tears and, like the suitcase, she wants to be carried, lifted into a sweeter present.

The wind hurls plastic bags up from the gutter and into the air. They fly like strange torn birds battered in a gale. Sheets of metal and bits of scaffolding fall from buildings onto the pavement. Mothers clutch their children and hurry home. The sky is the colour of dead fish. A man shuffles towards J.K., losing one of his slippers on the way. Beery breath, face much too close to hers, he whispers, 'Little woman, you have blood under your crotch,' and limps off into the swirling litter, a deranged prophet on the edge of the desert, walking into a swarm of locusts.

Lillian Strauss is planting red pokers in her garden. She

digs and digs. Sometimes she buries bottles in her garden, binliners full of bottles; vodka, whisky, gin, wine, and once as a joke, an SOS, tucked into a very special bottle of malt whisky, a secret note for someone to find in the future. 'My red hot pokers will bloom,' says Lillian Strauss, 'even in this bloody English soil.' And then she says to J.K., who hands her airmail letters so she can write to her sister in Singapore, 'You are a fool even though you are cunning and let me tell you this dear, you're hardly a matinee idol.'

'We're going to celebrate us in style J.K.'

Ebele orders two glasses of champagne and a plate of chips. Seven foot tall, in a suit and little red fez, he shows her his shoes. They are full of holes which he's stuffed with newspaper so that he creaks every time he walks. His wrists jingle with bracelets and they are in a bar in Oxford, surrounded by pale young couples eating salmon cakes; somewhere, bells are ringing, ringing through the Muzak and neat pastel blouses.

J.K. dips a chip into a puddle of ketchup, circling it round and round, smiling and looking away, and then looking straight into his black eyes – just as the Muzak drawls, 'Loooovin' you whah whah oooooooooooh.'

Ebele so big, the glass of champagne small in his hand, in his beautiful wide-palmed hand, bells ringing through his fingers, as he creaks and jingles, rocks backwards and forwards on his chair. He tells her he is a twin and how his twin brother died at birth, weighing two pounds. He, Ebele, weighed eleven pounds, and when the cord was cut, his brother died next to him, thin and sickly, and he a giant baby freak screaming on a slab in Sierre Leone, nose to nose with his starved dead brother.

'I felt a murderer y'know . . . for years after, so when I come to the West, first thing I do is look up a book to see how twins lie in the womb.' Ebele scoops up a handful of chips into his mouth and sips his champagne as if it was Guinness. 'I learn it was my mouth that was wrapped round the food supply innit.' His mouth is full of potato. 'Yeh. But the spirit of my twin brother is always with me. My baby brother. He is my lucky charm. When things go wrong in life I feel him with me. He is the boy who stops the planes I fly in from crashing.'

He claps his hands and the tassels on his fez shiver under the air conditioner.

'My hands feel so weak today. Look at my fingers. They're like spaghetti.'

Lillian Strauss has very thin earlobes and very thin lips. She has a biro in her hand and while she talks she doodles. The word Aristotle appears twice, and underneath it, a tortoise carrying a little red flower in its mouth. That's my insignia, she says. Like people have tartans I have tortoises. Her daughter laughs. In response to the laughter Lillian Strauss sketches a young girl holding a long eyelashed cat by the lead. Good night angel, she says, I can't drive you home because of my night blindness, and while she describes what it's like not to be able to see in the dark, J.K. looks at her mother's collection of tigers. They are arranged in little groups all over the house, striped heads and glassy eyes. Good night angel, she says, and her voice is panicked, breathless, just as it always is when she expresses affection. And then when everything seems okay, the words Good night angel, the puckering of lips to kiss, there is a sea change. Lillian Strauss says, you hate all my family, you don't want to know about my childhood in Singapore, you think I am a w.a.s.p., you ignore half your blood, and she begins to write in the same biro, a 'hymn of hate' to her daughter. The tigers look on. They look straight into J.K., yowling great cries into her heart.

There is a beast inside J.K. It is a mammoth, frozen in ice. It inhabits the colony of her interior and sometimes it stirs. While her mother writes the hymn of hate, she can feel it nudge its big ugly head against the ice. When her mother says, 'I love you so much,' it lies down again, and rests. 'You are going away again,' says Lillian Strauss. 'Good riddance.'

Starlings fill the sky. They circle a large whitewashed mansion with green shutters raised above the bay. Scarlet blooms grow in turquoise pots and trees bend in the breeze inside the walls of the garden. There is shade in that garden. And a hammock strung between lemon trees. There is health in that garden. Cool walls and birdsong. I'd get to look young in that place. I'd come home to rest in that place. I'd stop running, running through airports and railway stations, running through European cities looking for rooms and coffee and company and comfort. I would stop running away from this beast inside me. We would rest here and stop being frightened of each other.

Lillian Strauss has sold her house, sold her car ('four hundred pounds and it's yours'), sold a carpet, sold some silver cutlery, sold a bronze buddha, and moved to another suburb in London. She drinks a bottle of dry white wine at 11am and says to her daughter, I want a sea funeral, I want to be buried at sea. J.K. says, 'You've always liked the sea,' and gives her a clay tiger she has brought back from Spain. By 1pm her mother has finished the wine and is making scones. She is not a scone-making mother, but her mother made scones and she is trying to remember the recipe. She breaks lumps of butter into the flour and says, this is to let them breathe. The smell of scones cooking fills the kitchen and Lillian Strauss folds her arms over her soiled cardigan. She stares at her new tiger with dull eyes.

'I saw a park full of picnicking women yesterday afternoon. Young mothers and their children. They were picnicking on rugs and they were happy. I wanted to buy cherries and for us, you and me, to sit in the park and soak up the sun. I wanted to be as easy, as free and easy as those young mothers when I was a young mother.' She stops and her cheeks are burning. 'You have that horrible look on your face. You're always plotting.' Lillian Strauss is in one of her rages. She opens the oven and with her bare hands takes out the baking tin. Half the scones are sweet and half sizzle with melted cheese. She plunges her hands into them, tears them apart and throws them against the walls of the kitchen, her burnt hands writhing like snakes through the bone-white grass of her discontent.

Ebele brings J.K. one of his paintings for her birthday: an orange hand, its palm laced with henna, similar to Indian brides at weddings.

'Count the fingers,' he says.

'Six.' She smiles. 'Six orange fingers.'

'From your alien friend. They tell me I'm an alien at the airport.' He holds up his own fingers and tells J.K. to count them. One, two, three, four, five. She kisses his hand and then bursts into tears. Afterwards, as they walk in the park hand in hand, kicking piles of new mown grass into smaller piles, she tells him about her mother's blistered hands laced with sizzling cheese.

Lillian Strauss arrives at her daughter's house with a large tin of tomato soup and a black pudding sausage. The hem of her dress is held together with safety pins and her calves are scratched and bloody. Ha ha laughs Lillian Strauss. 'Just from the pins, dear. They come undone. What did you think they were?' Her cheeks are covered in a nerve rash. She thumps the black pudding on the table.

'Guess where I got the money to buy that.'

'Where?'

'I finished The Times crossword and won twenty quid.'

As they eat, Lillian Strauss points to the sausage pronged on her fork.

'What's this?'

'Meat.'

'No my dear. This is congealed blood.'

She puts it in her mouth and chomps with relish.

J.K. thinks about how much she loves her mother.

The panic of the raging beast. J.K. wears a summer dress the colour of the lemons she glimpsed in the walled garden. The colour she saw standing on the wrong side of paradise. Ebele stands behind her, plaiting her hair, brushing it, smoothing it down, weaving lemon ribbons into the braids.

Lillian Strauss takes a hammer and thrashes the ice tray. It is six o'clock, time for gin and tonic, a little bowl of peanuts, intimacies and brittle jokes. As the gin bottle empties, her hand tightens around her glass and laughter changes to melancholy. She begins to name all the cats she has owned and how each one of them died. 'I would have liked grandchildren to get sober for,' she weeps. One day J.K. gives her a present. Five papier mâché Chinese children. Round faces and black hair. There are blossoms painted on their clothes. Lillian Strauss arranges and rearranges them on a little straw mat, and J.K. observes that her mother has no half moons on her fingernails. Just as she is thinking this, Lillian Strauss flings one of the Chinese children to the floor and stamps on it with her square brown heels. 'What you need is a good kick up the arse. It's big enough.' The papier mâché baby lies crumpled on its stomach, cheek pressed into the carpet, and on the sole of Lillian Strauss's shoe, a little hand with three broken fingers.

Her mother points to the red hot pokers thrusting out of the stony soil of her garden. 'What do you think of my red garden, J.K.?' She leans over and strokes the stem. 'Your pink camellia, my dear, is for the cowardly.' The noise of the day fades.

When Lillian Strauss turns round to face the heat and silence behind her neck, she thinks her garden is on fire. And then she sees a mouth, a massive mouth, opening, opening, until it fills the whole of her eye, a quivering thing, standing in the blaze of her red hot pokers. She clenches her fist and thumps it into the mouth of the beast, alone with the child she created.

3

RE-IMAGINING THE STRANGER

He walks in although she has no memory of leaving the
door open. When she turns round to face him, he says,
'You are barefoot, you have one tooth missing, and you
are wearing a blue dress.'
   'Why do you always describe me?'
   Silence.
   'Your lips are cracked,' she says.

They are sitting together on the sofa. It is a warm night,
the heat of the day pouring in through the windows. He
takes out a wad of papers from his brown leather bag,
the bag travellers strap across their chests, a few essentials
packed with care. A book, a pen, a bottle of orange flower
water, a passport, a photo, a slab of chocolate, a wallet
heavy with foreign coins.
   'Currency,' he says.
   'I have been thinking about strangers,' she says.

He smiles and looks out of the window.

'How a stranger never belongs to a person or to a place. He can be an insider and an outsider at the same time.'

At that moment he puts his arms around her and her eyelash touches his cracked lips.

In bed they laze about for a while and then she climbs on top of him and says, 'Tell me about zones and frontiers.' He is kissing her shoulders, his lips are cool and he is saying, 'Um . . . ooh . . . there are um naked frontiers, take off your blue dress, and there are . . . um zones you can go into and zones you can't. Do you like being touched here or here?' She considers telling him where she was born, how old she is and what she does for a living.

'What's your name?' He puts his hand on her breast.

J.K.

She notices that this time (they have only been naked once together before) he shuts his eyes when last time he kept them open. She thinks he has his eyes shut because he is feeling something he did not feel before.

'You are naked apart from three silver bracelets,' he says.

'You're describing me again.'

'That's what strangers do. When they are in an unfamiliar place they describe it.'

'We are intimate strangers,' she says.

'Yes.'

They lie side by side, heads touching as the sky deepens and shops pull down their shutters.

'Do you want a glass of German champagne?'

'Was it a present?'

'Yes.'

'Why German?'

She shrugs.

'There's a story you're not telling me. That's how you keep someone a stranger.' And then, 'HEY you've got your shoes on. How can that be? You didn't have shoes on when I arrived. When did you put them on? We've been screwing and you've kept your shoes on!'

J.K. ties up the laces.

'These aren't just any shoes. They are made for walking long distances.'

She goes to fetch the champagne and, just as they open it, she lying across his back, there is a knock on the door. She says, 'That will be Zoya. A friend of mine.'

He is shy and surprised and puts his hand into the silver curls of his hair.

'But it's late . . . it's . . . '

'She has driven one hundred and ninety miles to see me.' J.K. pulls on her blue dress and leaves him naked in her bedroom.

Zoya wears little horn-rimmed spectacles (even though her eyes have perfect vision), a mantilla comb in her hair, and carries a small spherical black suitcase.

'It's the Doctor,' she guffaws. 'Where does it hurt?' Despite the humidity of this unnaturally warm night she also carries an overcoat. 'Got no love to keep me warm,' she says, and takes a pineapple out of her spherical black bag. She saws through the thick skin with a bread knife and sucks a ring of pulpy flesh. 'Those are the green plates you bought in Brixton market.' She points to a shelf above the fridge, catching the juice running down her chin with a cupped hand. The stranger walks in and strokes J.K.'s hair.

'You are wearing a blue dress, three silver bracelets, walking shoes, and you bought six green plates in Brixton market.'

Zoya adjusts her spectacles and lights a cigarette.

'Are you going to introduce us?'

Silence.

'This is J.K.,' he says.

The migrant stranger and the migrant Zoya sit together in another room while J.K. cooks. They begin to find oceans, motorways, railway lines, bus routes, facial characteristics, languages, bread, fruit, fish, jokes and musical instruments in common. When J.K. returns with plates of food and sits on a chair opposite them, she feels like a stranger.

The next evening, London is divided into two zones – by telephone. Inner and Outer, Central and Suburban. 071 and 081. The Post Office tower celebrates by spinning laser beams into the sky. Zoya and J.K. are walking down Charlotte Street, West London, looking for a place to eat lamb kebabs.

'How can I wo-rk when the sky's so blu-e. How can

I wo-rk like other wo-men do,' Zoya croons.

'I miss my family,' she says.

'You've never said that before.'

They sit down at a little café with tables and chairs sprawled on the pavement.

'It's the smell of lamb and heat.'

The waiter takes their order.

'I used to be able to speak Urdu as a child. Then this country beat it out of me. One night I got drunk in Berlin and remembered everything. I even remembered languages I didn't know I spoke! I remembered the house I grew up in, what part of the garden had shade and how I used to swill out the yard with buckets of cold water while my brothers played football. I was beside myself, babbling in tongues.' She takes off her fake spectacles, puts them into a red fake leather case and snaps it shut. 'And then I looked up into the cold grey eyes of the man sitting opposite me.'

'What man?'

'He told me he sold early Max Ernst etchings for millions of marks, all the time chasing a ribbon cut from the cheese he was eating across his plate. And then he talked about how he wanted to breed heavy horses . . . '

J.K. laughs.

'This was balm, J.K. I wanted to escape from the bloody pain sticking through my ribs, and heavy horses were just the thing. It could have been tortoises, stars, a list of rare ivy, buttonholes. He wore a crisp white shirt and citron cologne, and it was perfect, our difference was perfect. Like the Irish poet Patricia Scanlan who pushed all the grief of Belfast out of her head by writing lists of every sweet she bought as a kid at the local shop, I asked him the names and types of all the horses he wanted to breed.' She giggles. 'He told me he liked the idea of rubbing them down at five pm, when they were hot and panting! And I said things like hmmmm in-neresting, like I was a B-movie cop. I think he thought I was a headcase but he couldn't resist a rapt audience. We went our separate ways and I bought you a bottle of champagne to celebrate.'

'I drank it with the stranger.'

The waiter pours wine into Zoya's glass.

'This wine is the colour of health. It's not like other wines. This is medicine.' The waiter who is still standing by Zoya's side smiles and says something to her in a language J.K. doesn't understand.

'He says he will read our coffee cups.'

'You are flowers,' he says, and disappears into the heat and bustle of the restaurant.

'I miss my family too.' J.K. sucks a long green chilli.

'But they only live a tube ride away.'

'I know.'

Someone has taken a photograph of them. A sweet peppery perfume and a blinding flash. They look up into the golden teeth of a middle-aged man, trousers belted high over his paunch, carnation in his buttonhole.

'For you girls. Only five pounds.'

He waves the Polaroid through the air so that Zoya

and J.K. can see bits of themselves developing second by second.

'Are we here yet?'

Two women sit at a small slanting table covered with a plastic cloth. One has a glass of wine between her finger and thumb, hands resting on a red spectacle case. Her shoulders are turned in towards her friend and her lips make the shape of the word 'medicine'. The other has her legs crossed, bread in hand, chilli in the other. Her cheekbones are burnt from a day in the sun and her lips have just finished making the shape of the word 'family'.

'You are here.' The photographer points to each of them on the polaroid.

'We know we are here,' Zoya says.

He leans over and takes a toothpick from their table, tilts his head and works it into his golden teeth.

'Sometimes it's good to know where you are.'

The waiter lifts J.K.'s cup from the saucer.

'Why are you laughing?' she asks him.

'What's your name?'

'J.K.'

'I'll tell you one thing, J.K.'

'Okay.'

'You must give yourself a name.'

071 and 081 for London. The Post Office tower lights up and J.K. stares into her cup.

He walks in although she has no memory of leaving the door open. When she turns round to face him, he says, 'You are wearing jeans and a silk halterneck.'

'You have an airmail letter in your top pocket,' she says.

It is a cold night and it is beginning to rain. J.K. is thinking, I have just left the 071 zone to come back here, to these rooms, the books on my shelves, to the fruit in my bowl, and to this man. Is that what a home is? A place to invite strangers to? He is staring at the snakeskin buckles on her shoes.

'Y'know,' she says, and he turns his body towards her. 'My name is . . .'

His gaze shifts from her eyes to the radio behind her head. 'I like it that you're just called J.K.' When he lets his eyes meet hers she sees they are frightened.

' 'Bye,' she says.

He tangles his fingers with her fingers.

'What are you still doing here?'

He says, 'Start again. Why don't you take off your shoes and tell me who you are.'

'You like it that I'm just called J.K.'

'Tell me who you are so I can love you properly.'

She considers these words. There are eleven walnuts in a bowl by her feet and there are eleven words for her to consider. Tell-me-who-you-are-so-I-can-love-you-properly.

'You want me to be a stranger,' he says and, for the

first time ever, takes out a cigarette that smells of cloves, lights it and leans against the sofa.

'You even wear shoes in bed so you can walk away from me.'

She stands up, switches on the radio, and looks at him sitting there, in her home, too close, a coil of smoke above his silver head, airmail letter shivering in his top pocket. At that moment, the radio announces that war has broken out, and tanks are sliding through the ripples of the desert.

4

RIDING THE TIGER

An English Rose. The national emblem of England. The pink glow of the cheeks in health. Blooming Blushing Bright. But there are other roses. The rose of Jericho, of North Africa and Syria that curls into a ball in drought, the rose of Sharon that was probably a narcissus. The rose that covers the eyes of a corpse and the rosewater that scents lovers and sweetmeats.

Today Gregory told me on the telephone that he had Aids. We could both hear each other's TV, 60 miles between us, words like Saudi Arabia and The Allied Forces, and someone was knocking on Greg's door. 'Anyway,' he says, 'have a good holiday. Where are you going?'

J.K. walks through a 'BeachPark' built on volcanic rubble; swimming pool and pizza malls and discos by the pool and there's nothing else but creaking palms planted alongside white holiday bunkers, curve of beach and desperate sunset. Somewhere else, charismatic missiles glide above skyscrapers. The dark incense of burning date palms and eucalyptus trees fills the desert. Young boys in the uniforms of North America and Europe sleepwalk through the bones of abandoned cattle, unsettling stars, scorpions and a sun that makes them shiver and burn at the same time. Sand in their eyes, they circle oil wells, delirious under that enormous sky, while men older than themselves, also in uniform, murmur strategies, statistics, geographies, parables into their sleepy heads, make jokes, hand out sachets of ketchup and arrange funerals.

'It's all complex on this island,' says the supermarket senora as J.K., shipwrecked and solo, pays 340 pesetas for chorizo and bus timetables.

'What kind of place do you want?'

The words make her sad and nervous. She walks back to the English boys eating pizza at the complex. They stare at her, call out to her, throw rings of pepperoni and black olives at her as she ties the red ribbons of her espadrilles, maps and timetables open on her lap, and she, thinking about what kind of place she wants to be in, puts on her spectacles and stares back. She stares into their pale blue eyes, growing blonder and blonder in the sun (Malcolm X called them devils) and what she sees is struggling mothers, absent fathers, broken park benches under sagging grey skies, poor food eaten in small rooms, places she doesn't want to be in, places she has run away from in search of an imagined place, a place that is not this place, a place that is not that place, a place that is – a place that, like the words War and Peace, is perhaps just an idea. This is a very blue sky. Thin cats hiss into it. Cacti lament under it. Their golden spines shiver in the wind, and from the largest, most formidable of all, another scenario of struggle emerges.

From its prickly heart, to the soft waves not of the ocean, but of Muzak piped from the local boutique, Leon Trotsky emerges, shirt sleeves rolled up, tattered straw hat askew on his thinning hair. He says, 'Yes, I was indeed architect of an alternative world. But I was banished before I could make it. On the run. Carrying it with me. Heavy luggage, my dear, for a man who had to be nimble on his feet.' He considers the cactus next to him, prods it, takes out a small sharp blade and cuts a piece of its flesh. 'I grew very fond of cacti when I was in exile in Mexico. However, I missed the inspiration of heavy overcoats and I have always thought better in a fur hat.' He digs his hands into the black rubble. 'No good for potatoes. Yes I am the same Leon Trotsky who once wrote impromptu speeches on napkins in Moscow restaurants. In exile I felt the loss of newspapers very badly. But cacti suit my choleric temperament and, like me, they survive in harsh climates.' He smiles at a pale woman with straw-coloured hair and a T-shirt with ZAP POW MY WAY sprawled over her pastel breasts. She appears not to see him but her lips mouth the words CAR HIRE to some imaginary companion. 'Her nose is too short,' Trotsky observes. 'I like women with large noses who nonchalantly cross their legs.'

He stares at J.K., whose fingers are tangled in the red ribbons of her shoes. She sits on her unpacked bags, passport in her back pocket, counting pesetas. Trotsky screws up his eyes. 'Let us place you,' he says. 'You who are discontented, I can tell from the curl of your lips.' He looks around him, up at the sky, and then at his blackened fingernails. 'Why do you think you feel discontent more than those English boys eating Pizza Americano who will later get drunk and vomit in the swimming pool? Do you think they don't know they are poor miserable and needy? I was once an electrifier of weakening armies made up of just such boys.' He stops. Wheezes. Kicks the cat under his old brown leather boots.

'In the middle ages, these islands were visited only in the imagination. It took the map-maker Angelino Dulcert to record the actual sighting of the island and the humanist Boccaccio to describe his voyage of discovery in whatever ink and metaphor was available in 1341. Gold hunters and imperialists followed, and of course with all imperialists, slaves. 'You see,' says Trotsky, 'the island proved to be the most significant of the Atlantic archipelagos because the wind system linked them to the new world.'

He gestures towards the BeachPark Development where young couples carry plastic bags full of lager to their bunkers. 'I'd like a beer myself. A beer and my arms round my babe.' After a long silence he continues. 'We will jump six centuries or so ... where were we, 1341, let us consider 1936. Let us make it summer. Dust winds blow from the Sahara. Hungry goats scavenge for food. There is a drought and the wells have dried. A paunchy little man called Franco, once commander-in-chief of the Spanish army, meets up in the woods of La Esperanza in Tenerife with a few discontented officers. They promise to give him command of Spain's best troops, tough lean Moroccan mercenaries. They walk to a hotel – the patron is a sympathetic English man – and by the time they have finished their omelette and sherry, they have given him the foreign legion as well. By July 20th the islands are in Franco's hands and he thus begins to conduct his ideological orchestra with machine guns. Within hours, in the terrible heat of that summer, trade unionists, teachers, left wing politicians, writers and artists are imprisoned or murdered.' He stops. 'I'd like to shag that girl over there, blonde with muscles in her thighs.' He watches an English boy stick his tongue into her ear. 'Amorous vertigo in one of these BeachPark bunkers would really cheer me up.' He scratches his balls. 'I am undone. My hopes have beggared me.' Head bowed, he examines a small hole in his cuff. 'Just thought I'd give you some information,' Trotsky wheezes, 'it probably wasn't in the brochure, and by the way, I recommend the local banana.'

The virus is making sorties into my body. Today I coughed up green mucus into a bucket on my lap. My masseur, an East End boy called Spud, says when he massages me he can see another body floating above me, and that's the body he works on. He calls us Gregory One and Gregory Two.

And from the shivering centre of another volcanic cactus, transplanted from some other place, perhaps a happier place, into sunshine and shadow, into the gentle Muzak of the BeachPark, someone else emerges. Vladimir Illych Lenin. Ripped down from the bloodstained boulevards of Eastern Europe by his discontented children, Vladimir Illych, great orator with gimlet eyes, now a little shaky, blinks. He is not used to sunlight. He prefers burying his hands in his wife's mother's fur muff. Wiping the sweat from his brow with a handkerchief he says, 'The great sculptor Aronson was enraptured with my skull. He told me I resembled Socrates.' He smiles, bends down and cradles a small striped cat to his breast. 'Aronson told me, hands deep in the clay, that light emanated from my forehead, but my eyes glittering with irony and intelligence were not as protuberant as Socrates' eyes.' The cat purrs in his arms, its small paw catching his beard, and Vladimir, laughing, nudges it under the chin.

'My political wife, Nadia, she loved my brain, but my lovers liked my lips. Thick lips that give me a Tartar look. While I listened to Beethoven's Appassionata, played with kittens, read novels on couches, Turgenev my favourite but sometimes a little Hegel and Kant to keep me on my toes, nibbled cucumbers and made plans to hunt wild duck in autumn, I knew I could play my life in this way: admired for my lips, eat excellent goulash cooked by my faithful wife, enjoy lazy long games of chess, liaisons with admiring and full-breasted comrades, and write the odd book. I spent childhood summers in Kokushinkon reading Pushkin while my brother Sasha read Das Kapital – bought under the counter from a small second-hand book dealer. But I knew my destiny lay elsewhere! I would have to fight the seduction of Ludwig's Appassionata and ride the Russian Tiger.'

The cat, entranced by Vladimir Illych, falls into warm contented sleep.

'The day I sat my examination papers on Pushkin, the tsar sent my brother to the gallows. My mother's hair turned completely white and my sister Olga took to fainting at school . . . but I passed my exams brilliantly!' He stops, eyes settling on her chorizo, and asks how much it cost. 'And how many choices of sausage? Five? Yes, the people, they like to have a choice of sausage. The sum of my life's work undone by sausage. Remove it immediately, it offends me. To be deposed by a pig is not good for the morale.'

Vladimir continues. 'My brother, Sasha, argued that any philistine can theorise, but the revolutionary has to fight. The trouble with intellectuals is they are physically weak. They finish a debate, not because they have resolved something, but because they are tired. Stamina Stamina Stamina. Just raising his hand in a meeting is enough to make an intellectual die of exhaustion.' Stamina Stamina Stamina, the black volcanic rocks echo. 'Very big practical demands were made of the workers, but the intellectuals, they just want to screw each other and eat long lunches in cafés. It's the same the world over.' He stops again. 'There is something a little frivolous about the way you do your hair,' he says in a steely voice. 'I think you are under the influence of red wine when you should be under the influence of the workers' movement.' His mouth suddenly crumples and his small black eyes go moist. 'I HAD A DOG CALLED ZHENKA!' he screams to the sky. Zhenka Zhenka Zhenka, the black rocks wail. 'At twenty my brother Sasha pawned his gold medal to obtain nitric acid from Vilna, second-hand revolvers that did not fire, and explosives that were too weak. He died on the gallows because he engaged in political activity before he had clarified the principles on which it should be based. It is I who created the framework for well-elaborated principles. I had to put away my Pushkin and learn statistics. To cut the flesh and find the bone, lay in bare detail the economic connection between towns and villages, light and heavy industry, the working class and peasantry. What is that smell?' His nose twitches. 'Aaaah. It is your suntan oil.' He writes something down in a little notebook: 'What is the brand? Coconut with Vitamin F? Tested under Dermatological Control? Getest onder dermatologisch TOEZICHT!' Zicht Zicht Zicht. Seagulls cry above his furious head. Vladimir wipes his brow again and stuffs the handkerchief back into his breast pocket. 'I, Vladimir Illych, wrote "The Development Of Capitalism" in prison . . . FAKTOR FIFTEEN WATERPROOF . . .' the words seem to send him into despair. This time he howls. Proof Proof Proof, the black rocks howl back, and someone dives into the swimming pool. 'I wrote it for you . . . for them.' His hands gesture towards the pool which is now full of vomit and lager cans. He sighs, tickles the kitten's ears with his thumb, silent as he watches the English boys try to drown each other. He points his forefinger at J.K. 'Take your bags and leave at once. Tolstoy said when one travels, the first part of the journey is spent thinking about what one has left behind. The second half is spent anticipating what lies ahead.'

On a bus at the volcanic crossroads, away from the BeachPark, the bored driver plays with the buttons of his starched blue shirt. He says, you must go to Morro J, lays out his hand to conjure something beautiful for her there, and J.K. gives him her tortoiseshell fountain pen which he turns over in his hand, writes slowly in elaborate italics two words, *Pensione Omray*, and starts the empty bus. They're driving through desert dune to Puerto Rio for her bus connection, one hour to wait, three bags, two of books, one of clothes, Smith Corona 1936 typewriter in a pillowcase. He drops her at a small industrial port at the bottom of the hill, the sea whirling gases, a church, a cigarette kiosk, a local newspaper which has oil-drenched gulf birds on its cover: *Catastrofe ecologica en el golfo, contaminada por el crudo bombeado al mar*, and somewhere a ship's hooter shrieks while men gamble on the pavement. What is it, this blood that leaks from her mouth every day? Dark and morbid in the basin? Two red stars burn in J.K.'s cheeks. At the café by the bus station, small yellow butterflies knock against her bare arms.

'Spit it out.'

J.K. looks up into the eyes of a young black woman who takes out a tissue and says again, 'Spit it out.' The woman's daughter kicks her chair with her sandals and stares curiously at her. 'It's all right,' her mother says.

'It's all right to spit.' She calls out for some beers and an ice-cream for her daughter. A parrot cracking seeds on a perch nearby lifts its head and makes the sound 'Hooo Hoooo'. The daughter, shy, whispers 'Hoo' back and then looks away. 'Hoo,' the parrot calls to her, and despite herself, a little louder this time, her lips return, 'Hooo,' and then she hides her face in her arms.

'It's all right,' the woman says again.

'What's all right?' J.K. also hides her eyes.

'Nothing's all right,' the woman says, and they all watch the parrot.

'Last year I woke up feeling weird. I could hear birds singing, my body was warm, my fingers tingling and I was in the Promised Land! So I took a chair outside, outside my estate that is, and fell into the garbage. Yes, you might laugh now, but I could taste milk and honey in my mouth! The allotment was rustling with sugar cane! I reached out towards the cane and cut my hands on glass, my head was spinning and I walked the streets until I came to this synagogue and I ran inside it. There's a service going on and I shout UNITE! Everything stops and I shout again UNITE UNITE, so they called the police. Bloody Jews. I didn't say Fuck you, I said Unite.' She gives J.K. a tissue. 'That's better,' she says. 'Spit some more.'

'Hooo.'

'Hoooo,' her daughter begins, and then stops herself.

She sticks her tongue into the ice-cream and looks up at her mother.

'So the police put me in a cell, and in the cell are a lot

of blue blankets and blue was the colour of peace to me. So I thought, if I put these blankets up on the wall and over the door it's all going to be OK. But then they open the door and say they're taking me to the hospital. So I say, "You'll have to drag me there. I'm not finished with these blankets yet." And I sit on the blue blanket like it's a magic carpet and the police are pulling it, two on each end and me in the middle shouting UNITE UNITE, and they lash my arms behind my back and I end up in the hospital.'

'Hooo.'

'So nothing's all right is it?' Her finger prods J.K.'s hot arm.

A fat man sleeping on his guitar wakes up. He orders a plate of potatoes and chilli sauce, twisting the heavy ring on his fat finger round and round. He smiles at the child, teeth small brown stumps, and points to the parrot. 'Lauro,' he says and throws her one of his potatoes.

'A doctor in a white coat comes up to me and I knew I'd met him before somewhere. In a concentration camp, or he sold me as a slave, or he massacred my mother or deported my father or lynched my brother. I knew that man. I knew he was evil and could hurt me. So I screamed. That scream frightened me more than it frightened the doctor . . . I didn't know I had so much fear in me . . . and then I saw this woman, this black woman wheeling a trolley of tea, and she says, "You're frettin' darling." So I threw myself on her and stuck there like a leech and wouldn't let go and she walked around giving patients tea with me stuck to her, telling the doctors it was ALL RIGHT I was stuck to her.'

'And I was lost,' her daughter suddenly says.

'She bloody well was. But not as lost as her mother. They put this needle in my arse like I'm some kind of rhino . . . there was so much sleeping sickness in that injection I slept for three days. And then early one morning there's someone tapping my cheeks and I try to wake up . . . it's the tea trolley woman. She's got my clothes in a plastic bag and she's saying, "Get out of here. Get dressed and run for your life out of this hospital." And I see her eyes, they wake me up, I see too much in her bloody eyes, I see my own mother in her eyes and I get dressed and run for my life.'

'And I was lost,' her daughter says again.

'But I found you, didn't I?'

'Yes.' She hides her eyes again.

'So nothing's all right. Except I'm telling you this tale in the sunshine drinking a beer, and not in a nightie in Ward Two.' She points to a bus revving its engine. 'That's yours.'

Now she is rolling through mountains and red dust oases, beer and blood in her mouth, waving to the woman and her daughter. The fat man stretches out his arms and shouts, 'If you want my body you can have it!' The shape of the letters L and M cut into the sky, as if on a convict's cheek. L for lire, loony, levis, love. M for massacre, mint, molotov. J.K. spreads her hands over her lungs, palms warm and still, as if one part of the body can be sick and another heal it. She looks again at the scrap of paper. *Pensione Omray*. The bus shudders and stops. People get on carrying parcels wrapped in newspaper and string. Across the road an old woman thins out her tomato plants. On and on, from the North of the island to the South, herds of white goats and urban bunker developments, on past beaches of black sand, allotments growing tomatoes, solitary cafés, abandoned petrol stations and beat-up cars on the edge of crumbling cliffs.

Canaries twitter in small iron cages. Their master and tormentor, Omray, sweeps the floors of his pensione. Cigarette in his mouth, plastic sandals on his feet, he hums an old Elvis tune, soiled newspaper tucked into the pocket of his grey trousers. The jacket hangs over a chair at reception. He stands his broom against the wall and J.K. follows him to the chair, which he formally sits on, stares at her, lights another cigarette, asks for her passport which he flicks through, yellow fingernails tracing the outline of visas and the outline of her chin in the small photograph. His fingers move from Warsaw to her cheekbones, across Washington to her lips, eyes travelling over her luggage, especially the 1936 Corona in the pillowcase. Tired from his interrogation, he leans back in his chair and says, 'E-d-m-o-n-t-o-n.' The canaries beat their wings against the bars of their cages. 'When I am in England, I live on the edge of London. Dog's arse E-d-m-o-n-t-o-n. I prefer to live on the edge of life.'

He asks for some cash, counts it, locks it in a little steel box which he puts into a drawer, locks the drawer, and slowly, slowly, a smile parts his lips. 'Love me Tender, Love me True.' His keys jangle as he shows her a room with a little desk, an iron bed with a picture of a white horse above it. 'Chinese,' he says, and points to the shower which he walks lazily towards, swishing a plastic curtain around it with magnificent gesture: proud host, Omray, penned by a bus driver in italic ink, brought into being one Sunday morning at the crossroads. She has journeyed to him and his canaries who scream through the walls, and he, one hand tucked into the top of his trousers, screams back until they fall silent, sighs, smiles, reaches deep into his pocket and gives her a small yellow feather. 'Souvenir,' he says and closes the door very quietly, as if nervous he will awake the distressed birds.

J.K. wades into the thrash of the waves, deeper and deeper until she is floating with the gulls, looking out at the European couples walking the coastline. A sudden gust of wind blows white sand into their faces. For a moment, disorientated, the Europeans walk in zigzags across the dunes, displaced and dizzy, fists in their eyes.

A group of elderly Germans sit at a café, chairs arranged in a circle around the table, laughing and slapping their thighs. The oldest man of all, huge paunch hanging over his trousers, suddenly begins to choke, coughing and spluttering until water streams from his pale blue eyes and his steely spectacles fall to the ground. The more he chokes the more his friends laugh, clinking brandies and pointing to him, until, just as it seems he is going to breathe his last, he spits out, inch by inch, a long silver chain, pulling it from his throat with fat hands, mouth opening wider and wider as he pulls out a round silver watch. His wife claps her hands and roars, 'That was a good one! Better than the one you did in Munich,' and orders more brandies from the bewildered waiter. One of the men turns to J.K. and shouts, 'Why are you here?' The man who has just choked up the watch says, 'She's here to make her dreams come true.'

J.K. turns away from their pink smiling faces, her own face suddenly damp with tears. Why am I here? An Englishman sitting opposite her peels a boiled egg. He slams his eyelids down, blond lashes fluttering in some private excitement of his own.

As my body gets weaker, the things I most think about are pain and money. Perhaps my other body thinks about beauty and grace and how to measure value, but this one, my sweet, still has the same sort of fears people had in the iron age. Fear of the dark and certain kinds of animals. Things lurking under the sea, under my bed, inside my skin.

'In Beijing,' says the Englishman, 'the government had all the dogs shot.' He chews his egg slowly. 'I once shot my dog. She was called Ogre and I hated the way she looked at me. Too much. It was too much.' His teeth are flecked with egg yoke and he wipes his mouth on an old copy of the *Daily Mail*.

The sun is gentle, the ocean emerald, and somewhere windmills, a reddening creeper, a small garden with table and chair outside overlooking the sea. J.K. wants to sit there. Very badly. But she is not invited. She wants her own table and chair and garden and she hasn't got one. Insurmountable obstacles seem to deny her the possibility of ever claiming them. What does she have to do to get them? Why have some people got them and not her? To have a home is to have a biography. A narrative to refer to in years to come. There is a house in the garden. Turquoise paint peels off the front door which is half open. Sunlight pours through. It is self-possessed, inhabits itself to the stranger's eye with a particular kind of grace, has its own logic and order. Maps of the mind sprawl out and beyond the table and chairs standing in the small garden, spill into imagined scenarios of all kinds, but at this moment J.K. wants them to be part of her map. She wants to be able to point and say: these are the stones I dragged up and planted things between, these are the feathers and shells and cooking pots I collected, this is where I have placed them, this is the room I like most to sleep in, these are the paintings Ebele made for me.

How can she make the things she most wants happen? Not in dreams or sculpture or literature, but in bricks and mortar, with soil and seeds and water, in parliament, in the minds and hearts of other people? Who is the citizen sitting with her on those chairs in that garden? What does it mean to be named a citizen? This citizen is prone to violence and that citizen is prone to barbecues in Hertfordshire. This citizen has spent all her historical time surviving, getting wrecked in clubs, murder in her heart, cocaine up her nose, she rises from the eternal, dreary, fetishised flames of her own anger and says to that citizen, so COME ON then, tell me about tolerance, moderation, your neighbourhood, your schools, tact, good manners, tell me about your Gods AND all the wars you fought in. I'll tell you about my neighbourhood, schools, taxis, clothes, ecstasy, drag queens, any number of sad corrugated sunsets AND all the wars I fought in. Tell me about this world and how to be well inside it?

Today Gregory says there are worlds floating in his bloodstream. Sometimes they make him feel beautiful and delirious.

The arrogance of metaphor when facts save people's lives. The succour of metaphor when facts inadequately describe people's lives. The bravado of T who wore crazy jewels and made sweet wine from berries growing on the banks of railway lines. Abandoned with her small daughter in a high-rise in Bethnal Green, but growing her up good with fruit and books. And C whose twin sister suicided herself whilst swimming in a river one tearful summer. How she decided not to come up again, to put her head under and disappear, and C forever hallucinating her sister in a yellow dress, drinking coffee, eating bread, saying stupid things like Continental Blend and Yardley – as if the century had taken away her language and all she had left were brand names to describe herself. Her breasts dark circles under the yellow dress as they hoiked her out of the river, eternally hoiking woman out of ponds and lakes and oceans. She just wades in and goes under, all furtive and furious in useless protest, hoiked up by some geezer in wellies, leaving her sister to mourn and hate her.

And M who never travelled anywhere, except to the liquor store and back and back again and back and back again, who wrote poems and sent them to her, terse with the fear of humiliation, literary references and cryptic asides. How is it that M, alone and broke, drinking away her intelligence in front of the television, imagines her constituency to be professorial gents in corduroy with Anglo-Saxon beards and wives who sacrificed their lives to nurture the sensitive interpretive twitches of their literary husbands – and she, M, describing her life in language that doesn't fit her, that is to say, adopting the puns, tone and form of those whose lives are cosier than her own: a regular salary, children grown up by someone else – never read the world but a dab hand at sonnets, sonatas, Elizabethan musical instruments and logical reasoned argument.

Mega-Star! Mega-star! The Englishman who shot his dog shuffles through the market, chanting, a small jar of Nescafé under his arm. A turkey escapes from its cage and runs towards him, gobbling leaves and flapping its wings against his flip-flops. He kicks it away, making turkey

45

noises in his throat, grabbing feathers and sticking them into his hair. COME ON COME ON COME ON. It runs back to him. YALALALALAYALALALALAYALALA. The only sound that can be heard above his warcry and the writhing turkey are the words THE ALLIED FORCES.

Hurrah! I've got pneumonia. I've been blitzed! I'm a goner! I'm all technology and biology! Half alive, half dead. I'm God. A machine measures my heartbeats. Five drips poke into my body. My mother sent me some tartan socks and a peculiar card saying that when I was born she couldn't decide whether to call me Klaus or Gregory. This coincided with an old friend changing his name from Eric to Gus. When people suddenly out of the blue change their name, I always think they've been visited by strange men in space ships. Out of the blue. Where is the blue? The Blue is somewhere. Where are you?

What cultural violence made M's poems so boring? Why did she need the approval of a canon that would never invite her bad-tempered brilliance, politics, poverty and ungainly female form to their dinner table?

Does M exist?
What proof does she have?
When did she become a person?
When did she cease to become a person?
What kind of language is going to (re)create her?

In troubled dreams the white 'Chinese' horse on her hotel

wall gallops across J.K.'s stomach and tells her in strange whispers that he will return. His breath is warm and wet, sometimes he speaks in Mandarin, sometimes in Spanish, and he does return, this time to say in strange hieroglyphics made from ice: We return to homelands and find they are a hallucination. We return to our mothers and fathers and find they are not the people we thought they were. We return to our street and find it has been re-named. We return to our cities and find they have been rebuilt. We return to our lovers and find they are elsewhere even when they lie in our bed. We return to our people and find they have been massacred and we were not there to defend them. The redemptive homeland, she is a joker, she runs away bells ringing on her toes, you chase her at your peril because she will appear disguised as something else and you will be chasing her all your life, watching her fickle back turn corners. What are you returning to, J.K.? What is your name, what are your voices, and most importantly, what are your actions? What use is the heart turned inwards? That is a lonely home, it knows each crack in the ceiling and every stain on the carpet. It must gallop outwards into the wilderness and perhaps even die there. Come out to play J.K.

She wakes to find Omray standing above her bed, cigarette glowing in the dark and canaries screaming in the corridor. 'I've bought you some more souvenirs,' he says, and drops a handful of yellow feathers onto her belly. She packs her bags, slams a roll of pesetas into his sleepy grabbing hands and walks out to the bus drivers' bar for churro and cortado, three bags, one Smith Corona 1936 typewriter in a pillowcase, and the breathy syllables of the horse tattooed on her face.

A prostitute with bruised elbows sits on a high stool drinking warm milk, a yellow plastic flower about to fall from her thin black hair. She has let her shoes drop to the floor and her ankles nudge each other as she avoids the eye of one particular bus driver who drinks half a pint of lager nearby. J.K. sits next to her, bags by her feet. The prostitute glances at the pillowcase and then at J.K., who smiles as the patron brings her a plate heaped with churro and a small coffee. She likes mornings. The beaches are empty, streets are being cleaned, and people have not yet summoned their meanest selves to pull them through the day. She last ate churro with a lover two years ago in Southern Spain. He dipped the sausage-like thing into hot chocolate and said, 'I love the blue rhinestones in your ears, by the way.' Usually a man of few words, an occasional joke and wry smile, observing her laughter but keeping his own inside him, that morning he talked and talked. Had she seen this and had she read that and what about hiring bicycles and heading off to a village famous for its honey and how brown her legs were getting and how much he liked the cool of marble floors and why did she cry that day in Lisbon and how he painted with coffee as a child in Argentina because his family were poor and could not afford to buy him paint and how his first wife died in a car crash leaving their ten-year-old daughter unable to sleep at night for fear of waking up and no one being alive and how she speaks French, German and Spanish and says she wants justice in three languages, how he planted English yellow flowers, what are they called, daffodils, in two old kettles and, eventually – I love you J.K. – the words spoken for the first time, up to now always avoided, loud and brave over a plate of churro, and J.K., blue rhinestones in her ears, silent, receiving the words and not returning them. She bends down and picks up the yellow plastic flower that has fallen from the prostitute's hair that reminds her of those yellow flowers planted with love so long ago in two old kettles. Gracias, the prostitute says, and the pin-ball machine in the corner whirs in the black pools of her eyes.

Stretched out on a sand dune high above the sea, cheek pressed into the sand, J.K. watches the sun slip bloodily into the purple ocean, radio tuned for news and the sky darkening as hours slip by. Strange voices leak through as she stares out across the horizon, shivering in a thin dress under the stars:

ISRAEL, THE ALLIANCE, SAUDI ARABIA, 2,000 SORTIES, 5,000 CIVILIAN DEATHS, DENYING THE ENEMY AN INFRA-STRUCTURE, LIMBS OF WOUNDED CHILDREN AMPUTATED IN CANDLELIGHT, ROCKEYE CLUSTER BOMBS, NEEDLE SHARP FRAGMENTS

Here, it is night. Cafés by the sea are busy. Hostile fatigued waiters carry trays laden with ice creams and beer and escalopes to bronzed men and women. Local fishermen, shoulders tense, stand against walls flicking worry beads, shuffling sandalled feet, smoking cigarettes, eyes on the ground, listening to the radio. Still and bowed. There are not enough fish in the ocean for that gut appetite. Tonight the Europeans are hungry, they want to be filled up. Fists bigger than local chickens, they complain about hire car firms in between mouthfuls ... ALLIED FORCES, WE ARE THE ONLY NATION ON THIS EARTH ... and J.K., lonesome cheek pressed in the dark, watches their shadowy arms lift glasses and forks like giant ghosts from a world that is too familiar. It is possible, though, that it is she who is the ghost, invisible, disenfranchised, the fragile daughter of colonial wars, one brown hand poking through the belly of Western Europe, the other wrapped around a bottle of malt whisky.

J.K. on a sand dune lit by stars and light from fishing boats on the tremendous ocean.

One winter she ran away to the flat marshland of South East England and lay on the pebble beach in the rain, sea lashing, just lying there for two whole days and nights. Three months later, feeling better, she unpacked her bag full of maps, any maps, ancient maps of China, maps drawn in 1310 by the Byzantine monk Maximos Planudes in response to the writings of astronomer Ptolemy, ink etchings of maps impressed on small clay tablets from Babylonia in 500BC where the universe floats on the sea in the form of a disc. Manuscripts which divided the earth into seven parts of the body: backbone, diaphragm, legs, feet, throat, rectum, head and face. She studied how the vocabulary of form changed with conquest, how the geography of speech and desire have all known invasions, plunderings, struggles and disguises. There in that marshland so bleak she could only look at it in parts, the horizon a long white scar, she thought about the instruments of early science used for surveying, measuring and mapping the world. The lenses, microscopes and telescopes that helped the subject get nearer or further away from her object of study, that led her through unknown worlds to the theatre of the galaxies. The further her mind wandered, the more curious she became about inscribing experience and information: if maps correspond to reality as seen at a particular time, what happens if she observes a number of realities at the same time? The word 'Perhaps', which could be a route to possible worlds, but used in a certain way becomes the route to a single conclusion. Unlike the word 'If', which implies the discovery of possible universes, by making them.

Her brother sent her a book. The postwoman asked her to sign for it but she had forgotten her name and didn't know how to tell her. And then she saw J.K. written by her brother's hand on the parcel and copied it letter by letter as if she had just learnt her ABC, and the postwoman was gentle, helping her out, laughing in the right way, so she offered her coffee and for six weeks took sea walks with her, made pancakes with her, let her brush the knots out of her hair, just glad she was there, finding ways to keep her warm, stopping drafts that raged through doors and windows. A loner with intelligent fingers.

Loners are the opposition. Pensive, thoughtful and furious, marooned with stories that need to be spoken out
loud and no one to listen, curries to be cooked and
no one to taste, days and days of traffic signals to be
manoeuvred and no one to congratulate, except other
loners: they find each other because like all good
maps there are familiar signs that lead the way. The
loner who both observes and creates worlds necessarily
speaks with many tongues. It is with these tongues that
she explores the contours of the centre and the margins,
the signs for somewhere and elsewhere and here and
now.

J.K. stranded on a sand dune between a war, three
bags and one pillowcase.

*Rockahulla! Blind. Almost blind. My head is full of
dizzy blond Muzak. The kind you get while waiting to
be connected on telephones. Oh yes! No Bach chorale
for me. My head is full of form. Donald Duck! Mickey
Mouse! I've been invaded by an army of Disney pets
when I should be at my most profound. Are these my
inheritance? Fear of death comes and goes. It's life's the
edgy thing. You always wanted a garden, J.K. That's easy.
Happy composting. Glorious growth. Glorious everything.
What can you see?*

J.K. sees the owner of the small sea shack on the cliff; table and chair outside, the table and chair she longs for, cacti and bush of herbs, boat on its side and tottering TV aerial on the tin and tile roof, palms rattling in the cold wind. A woman in a turmeric dress, bare legs and strong shoes waters her plants, looking out to sea while water spills on cacti. Someone comes out of the doorway carrying two bottles of beer, perhaps her mother, silver plait coiled around her brown head, pointing to a thirsty shrub. The turmeric woman is lost in some reverie of her own, ignoring commands to water this and water that, stopping now and again to sip beer or examine the broad leaf of a succulent. How did she come to be there? Who is she? J.K. sees her own mother as observed by herself at five years old, pins in her mouth, french pleating her hair – it must have been early 1960s – watching her dress in the mornings, catching the thrill of her presence in gardens or leaning against a car. When she was J.K.'s age she had three children, had been married twice and was now alone, struggling with debtors demanding money she hadn't got. They would eat bread and apricot jam one two three days, and then on payday, steak, a new sack of oranges to suck in the shade. J.K. barefoot, lying on her stomach peeling oranges. Reading in the long grass. She is frightened and she is ashamed. Sometimes she cries and no one knows. Where is her father? She is nine years old and she knows that sometimes people are tortured. Are grown-ups cruel, then? She looks at them in a new sort of way and when they catch her eye she immediately smiles in case they know she knows they are capable of doing cruel things.

Who is going to love her enough to make her brave?

Her mother wore false eyelashes sometimes and lipstick and listened to classical music, but also blues, drank brandy, smoked cigarettes in a holder which she lost often and they had to search the house top to bottom while she went mad until, victory between their teeth, one of the children would find it and she would kiss them all over, laughing again. J.K. remembers thinking her mother was lovely and beautiful. She was allured to her, pulled to her, zipped the back of dresses for her, wanted her, tried on shoes at the bottom of her wardrobe clandestinely, especially in love with a purple patent pair with straps which seemed to promise a glamorous future, unknown worlds that J.K., five years old, glimpsed as she did up the buckles. Every kiss was a treat, Sundays a treat, tickling the soles of her mother's feet while she read newspapers in the sun, coffee and slacks with zips at the side. Love is no maiden in silk. She is a monster who bellows, ugly and wounded. And her children are ugly and wounded too.

J.K. picks up her bags and resolves to find a place to stay.

In the chapel of the local monastery three monks gather around a painting of the Madonna, one perfect incandescent breast exposed to feed the child in her arms, nipple erect and moist. J.K. sits outside the chapel in the shade of an ancient tree, its trunk marked with three white circles of paint. She closes her eyes.

'Just ships passing in the night.'

The green glassy eyes of the Englishman settle on her breasts. He moves nearer, immaculately manicured finger-nails flickering across his trousers. J.K. thinks, yes, I am sitting under a tree marked for death. I must ask the monks something. What is it? Oh yes, when they think about woman what do they feel?

'I'm a mega-star,' the Englishman says, puts two fingers to his forehead, shouts 'bang bang' and collapses at her feet.

Today Gregory died. Slipped out of this century. A few days earlier he said, here's a picture of us, J.K.: we're talking about places we feel happy in and people we feel happy with, about our ordinary everyday lives and the planning of things to look forward to. God is Dead. Long live lager!

'Bang, bang.' The Englishman falls onto the crackling leaves by the bench. Everytime he makes the sound 'bang bang' in his throat, he dies again, in slow motion, mouth open, miming some terrible agony of his own.

J.K. is looking for a piece of string to tie up her suitcase which has split. Gulls cry above the glittering ocean. Grief is an inflammation. She spits it up bloodily, unhealthily, stupidly. She wants it to go away, but it won't let go. She can taste it and see it and she has to spit it out. Here, the fishermen's nails are crammed with fishguts, tourists translate menus, and dogs sleep under cars. There, a friend has died, it is a cold winter and trains have stopped running.

Somewhere else, strewn across the desert, corpses, charred limbs, the odd shaving brush, a microwave, a mirror and one broken bottle of rose-scented cologne soaking into ripples of sand.

In Washington the currency is dollars, the bread yeasted, breakfast waffles and maple syrup, coffee filtered and decaffeinated, golf is being played on slopes of green grass and yellow ribbons hung on taxis. In Baghdad, the currency is Dinars, the bread unleavened, breakfast goat's cheese, coffee flavoured with cardamom, foreheads scented. Mustansiriya college in the centre is the oldest university in the world, crops are rice, vegetables, maize, millet, sugarcane, pulses and dates.

Do we exist?
What proof do we have?
When did we become a people?
When did we stop becoming a people?
What kind of language will (re)create us?
It is possible that classic rules of form and structure do not fit this experience of existing and not existing at the same time.

J.K. watches a storm rage into the crimson afternoon.

The sky is electric. Rain whips her bare arms and legs. Dustbins are hauled into the air, caught on the wind's curve. Bags and pillowcase unpacked for a while, toothbrush, perfume, books, a little pile of yellow feathers, J.K. knows she too is caught in the wind. She is Europe's eerie child, and she is part of the storm.

5

BOOK OF THE OPEN MOUTH

Rain lashes against car windows. Her favourite dress lies in a heap on the floor covered in candlewax. The white wax against black velvet looks like a fierce livid scar. The scar above her eyebrow makes the shape of a K which is the second letter of her name. J.K. shuts her eyes.

H arrives. They have met once before, briefly in a train where she felt the brooding and bemused attention of someone staring at the wet black fur of her Russian hat (it had been raining), which she placed on her lap, lightly caressing its fur as the train rattled through the smoke of belching industrial chimneys. When at last he spoke, it was to conjure a picture for her. 'Your hat makes me think of the time I thought I was going to die. I was standing on a jetty. There was a raging wind and a huge wave of white froth seemed to curve above my head. I thought I was drowning. At that moment I looked down and saw a black kitten sleeping on the wooden boards.' He waited for her to say something or ask him something and when she did not he said, 'The white smoke from the chimney reminded me.' J.K. guessed from his voice that he was German, and another image of smoke raging from chimneys presented itself to her.

Now, as he walks through her front door, gift in his hand, he comments on the pleat in the sleeve of her black velvet dress, the books on her shelves and the thick ivory candles flickering in two heavy Ukrainian candleholders. J.K. pours rum into two long stemmed glasses. She is ill. Flu is streaming through her, a virus, it is the decade of virus, and H, who offers her his handkerchief, is in a maverick mood.

Three days earlier, as she shut her front door, unlit cigarette in one hand, box of matches in the other, and started to walk down the stairs, a short man in his thirties walked up the stairs. They collided and he quickly shoved his hand inside her skirt. In the fight that followed with this stranger on the stone stairway, he somehow manoeuvred her onto his shoulders so that she, still with the matches in her hand, was on top of him, looking down at the frizzy blond curls of his hair. He was struggling with her weight and at the same time running his hand up her thighs. Suddenly she knew what to do. She lit a match and set fire to his hair.

After they have eaten, H turns his chair towards her

and says, 'You look like a matador. You would fight small bulls, though. The sort you see running wild in the Camargue.' She lights his cigarette and asks him what his accent is. 'German,' he says. 'I like cold winters.' They drink more rum and she unwraps the gift he has brought her. It is a small packet of wild rice.

'Wars were fought over that rice.' H strokes the grey suede of her shoe. 'In fact it is not rice at all. It is a black seed that grows into aquatic grass in certain parts of North America.' As they dance across the curved room, tasting the rum on each other's lips, her hand pressed into the back of his neck, his hand pressed over her heart, which is beating fast, something salty mingles with the taste of rum. It is her tears, streaming again, and he moves his hand from her heart to her cheek. After a while she says, 'What kind of places would the trains journey through in a united Germany?' His fingers, now wet from her tears, draw a map across her cheek: 'Erfurt-Leipzig-Potsdam-Berlin.'

She lit a match and set fire to his hair. The blond stranger on the stone stairway began to burn, his frizzy hair in flames, the palms of his hands slapping upwards, anywhere, her calves, her knees, still holding onto her, until he got desperate and began to dig his sharp nails into her stomach and finally into her forehead, making the shape of a letter K.

'What shall we do about your flu?' H whispers as they dance into the flickering light of the candles. 'Tell me about Erfurt,' she says. His pale eyes settle on a painting behind her. Two vultures hover over a cream satin slipper, a languid red rose on its buckle. Next to it, a thin bamboo stick pokes out of a pot filled with soil, thin strips of shiny paper, gold, purple, green, glued to it so that if whirled it creates an arc of light and colour. 'In this room you have made yourself a world that pleases you,' he says. 'In Erfurt there is a cathedral. The houses are covered in soot and the air smells of coal smoke. There is also a theatre and . . . ' he smiles . . . 'good ice-cream.' She follows his gaze as they dance, reading book titles as if they were new to her, and when they kiss under a small book called *Undocumented Lives: Britain's Unauthorized Migrant Workers* she says, 'Well, I think we should go to bed and drink more rum.' He smiles, sticks his finger into her mouth and says, 'Um . . . you see . . . you are quite lovely, but the thing is I have another involvement and I don't want to lie to her.' They dance in silence, this time his hands in her hair and she says, 'So tell me about Berlin.'

'Berlin,' he begins, and then stops. 'Berlin is where I was born. Erfurt is where my . . . my . . . companion was born. She comes to Berlin to buy lipstick.'

'And to see you,' J.K. interrupts.

'Yes.' The vultures and satin slipper seem to fascinate his eye. He dances her closer to the image and studies it.

'The bird has a snake in its mouth,' he says.

'Were you standing with her on the jetty when you thought you were drowning?'

'Yes.'

'Why did you say I and not We?' she asks as they dance on and she untangles his hands from her hair, holding onto his hand though, both warm from the rum and he pressing against her velvet hips, ambivalent and desiring, his pale eyes somewhere else.

'Because it was my thought,' he says. 'The white smoke and your black fur hat. The white wave and the black kitten.' He takes his finger out of her mouth and presses it against the scar on her forehead. 'K,' he murmurs. At that moment his elbow nudges the Ukrainian candleholder.

She lit a match and set fire to his hair. At last she managed to jump off his shoulders, calves and forehead bleeding, and ran down the stairs leaving him folded over himself, slapping at his blond head with blistered hands. And then, he spoke.

'And because,' he looks away, 'I want to beam love into you.'

As hot white wax trickles down her black velvet breasts, J.K. sees the packet of wild rice lying on the table, a delicacy, a frivolous gift, and pulls in the following memory:

A woman holds up a queue of impatient (West) Berlin office workers one lunchtime in a supermarket while her groceries are cashed up at the till. At the other end of the supermarket (East) Germans queue for shopping trolleys because a sign tells them to. The shop is crowded with people pushing empty trolleys, a can of beer in one, a box of washing powder in another, two bananas and a can of

62

Pepsi in another. No one can move. There are skid-marks on the lino from the wheels. An old man reads the label of a small carton of cream, broken shoes tapping against the beat of Muzak spilling through the speakers. He puts the cream very carefully into his trolley, walks to the cashpoint, stops, bends down to pick it up and read it again. Eventually the woman turns round to face the office workers who are having to dodge the trolleys squeaking past them. They do not have trolleys. They carry their groceries in their hands and have currency ready to pay and leave.

'I queued for food for twenty years, you can queue for twenty minutes. Look! My mouth is open.'

They shout back at her, call her a White Turk, and she becomes quiet as she takes them in, their perfumes, shoes, briefcases, watches, cuff-links, haircuts, jewellery. 'Are you the new world I've been promised?'

J.K. stares into H's pale eyes.

'It's not a good idea to stick your finger into the mouth of a hungry woman.'

'Who is more predatory ... the satin slipper with rosebuds on its buckle ... or the bird above it?' he says pointing to the painting.

And then:

'Tell me how you hurt your forehead?'

His hair was on fire and then his mouth opened and words poured out. 'I gotta plate inside my head. Some cocksucker cracked my head. Only wanted an aspirin.

Looking for an aspirin. Need an aspirin. I buried the dead in Bucharest, Miss. Threw apples on the graves, six foot under the snow. A HAPPY NEW YEAR IN LIBERTY! If you've not got an aspirin, can you spare a piece of cheese?

'My companion and I are together because we are frightened to be alone.' His fingers search for the zip of her black velvet dress. 'But we are alone. I live half my life, pretending I am full up.' Outside, bins topple as drenched city cats search for food.

'I am touched,' J.K. says to her mother, Lillian Strauss. 'I am touched by H in every way.'

'Give some more form to the object of your affection,' Lillian Strauss says softly, trying not to smile.

'How do you mean?'

'What's his name?'

'I'll tell you when you're sober.'

Lillian Strauss's hands tighten around her glass of scotch. 'You self-righteous pious little shit.' Her eyes go dim. 'Why do you have to ruin everything?' They sit in silence. Lillian Strauss takes a small sip from her glass and purses her lips. J.K. looks out of the window.

'I enjoyed Gregory's funeral.' Her mother takes another sip.

'Thank you for inviting me. I liked his mother. She said if she'd called him Klaus and not Gregory he might not have got Aids. She's a bit weird isn't she? We're having supper together on Tuesday.' She looks at her daughter whose eyes are glued to the window.

'If I'd known I was going to blubber, I would have taken a tissue.' Lillian Strauss stands up. Walks to the sink and pours her drink into it.

'Bloody good stuff to waste,' she says, slamming down the empty glass.

'Mom,' J.K. says.

'Don't call me mom. And don't ever have children. They'll just end up hating you. That's what happens to parents. Their children hate them.'

'Let's have a baby,' H says to J.K. His hand rests on her belly. It is summer. A small aeroplane hums above them. Her camellia has flowered again, another pink bud opening in the petrol winds of the city. She looks around her room; a little saucer full of yellow canary feathers, pebbles, postcards, a bag full of coins, an address book, a white bowl on a stand, a photograph of Gregory, a cashew nut in its shell – not unlike a foetus, a poster of a man with a dragonfly taped to his forehead, a green ribbon, the letters XYZ scrawled on the back of an envelope in felt pen, a picture of an orange hand with six fingers, 'ALIEN' written underneath it, and a 1936 Smith Corona typewriter. J.K. feels panic rise in her chest. The same panic she always feels when arriving at a new place. She is in a new place. She is in H's arms and the aeroplane nearly drowned out his words, but she heard them and he is waiting for her reply.

This is a frightening place. His hand on her belly. More frightening than walking the city late at night, alone, in clothes that make running away difficult. Than the crazy gaze of bureaucrats in uniform, thin youths with knives, the violent hands of a commuter in pinstripes.

'I hope', H says, 'that when I touch you, you can feel everything I feel for you.'

Mother. The word is full of pain and rage and love. Her children play in small city parks. Cut their feet on glass. Howl into pillows. That's what children do. They howl into pillows. Howl for justice, for beans, for God, for love.
  'I'll think about it.'
  After a while he says:
  'I'll drive you to the airport.'

She is walking past a cement factory, straw hat on her head, books under her arm. J.K. knows she will have to collect ten one hundred peseta coins to phone H from a call box. She knows she will have to find out what the international dialling code is and she will have to find a voice to talk to him with. She could say, why are you there and I'm here? She could say, I'm learning the language, I'm brown and strong, the scar on my forehead is completely gone, every day I dive into the sea and every day is full of you. And then she remembers the eyes of a woman in her early fifties irrigating her garden in Southern Europe, drinking a glass of home-made lemonade at the end of the day after she had scrubbed the soil from her fingernails. What was that look in her eyes? Betrayal. J.K. knew she had been betrayed. Utterly. And the woman knew J.K. knew so she felt humiliated and when their eyes met J.K. had to pretend not to know. She had to find a way of meeting her eye dispossessed of knowledge. What does J.K. know? She knows that no one is innocent. Only the privileged and sentimental can afford to be unknowing.

J.K. is guilty. She buys the black-haired waitress at the local bar a beer and asks her to describe her room and all the things in it. And who do you love? And how do you survive on your wages? And how is your life different from your mother's life? And then, much later, she asks her for some hundred peseta coins for the telephone box.

Telefonica-dialling codes:
COLUMBIA 07-57
BRASIL 07-55
EMIRATOS ARABES-UNIDOS 07-971
CHILE 07-38
YUGOSLAVIA 07-38
INDONESIA 07-62

J.K. studies codes. A code is a collection of laws. A system of rules and regulations, of signals and symbols. So now, as she drops the coins into the steel slot of the telephone, she is thinking about rules, signals and symbols. H says, 'Is that you J.K?'

My precious.

My sweet.

My darling.

My lovely.

What is German for 'The twentieth century'?

Das zwanzigste Jahrhundert.

And how do you say 'Enigma'?

Enigma.

J.K. has two coins left. She rolls them into the slot.

And how do you say, 'The Open Mouth'?

Der Offene Mund.

6

SWALLOWING GEOGRAPHY

She is the wanderer, bum, émigré, refugee, deportee, rambler, strolling player. Sometimes she would like to be a settler, but curiosity, grief and disaffection forbid it. She is however in love with the settler X, he being all that she is not.

Today she found two birds' eggs, pale blue, one in a field and one on the pavement of a city. She buys an envelope in a post office so she can send them to a friend who will appreciate them. The clerk is intrigued that she is wrapping two small cracked bloody eggs in a sheet of white tissue paper and putting them into an envelope. This same envelope will be stamped, inscribed with the name of the place she has posted it from. The receiver will now be able to imagine the sender in that place and make a picture of it in her mind's eye.

The wanderer Y is not without purpose, but the purpose is not wholly revealed. Sometimes she imagines the layout of land before she has actually seen it. Instead of following a map, she has made a map. Sometimes she is visible and sometimes she is invisible. This is not because she is a ghost or a mystic, but because some people want to see her and some people do not. The word absence suggests non-presence, loss, being nowhere, non-appearance, lacking. That's what the Turkish worker on a German tram told her, fanning out his hands for her so she could see the offal under his fingernails. 'This is the liver of a cow,' he said. 'We at the slaughterhouse carry the inner organs of beasts, carry their bodies on bits of our bodies.'

When she meets a stranger and they tell each other stories, she notes that it is always the people she leaves out of the stories that interest the stranger most. If she talks about her brother, sister and father, the stranger wants to know all about her mother. Therefore she learns that absence is often more interesting than presence.

Although she is walking forwards, one foot in front of the other, she is also walking backwards. This is because she is thinking of her past. Beginnings and endings curl into each other like a snake with its tail in its mouth. There was a man who wept and said, 'I can't remember myself. I see and recognise myself in the bodies and voices of other people. I know that we once worshipped in the same temple. I know that the same priests blessed us with basil leaves and water. I know we fought the same revolutions, told the same jokes and went to the same schools. I have been described somewhere but I don't know where to find myself.'

The settler X kisses the wanderer Y on lips that are cracked from wind, and says, this is the statue of my local poet, engineer, architect, painter, banker, philanthropist, scientist, mayor, and here, a statue of the local martyr. He takes her by the arm and points; this is my park, my shop, my dentist, doctor and baker. She is eating a burger and chips. Always be ready to eat a burger and chips, a Czech refugee said once, in a television studio. All the while he had a kitten on his lap and he stroked it. He said the cat wandered into the studio and he wanted to be filmed stroking it whilst talking about exile, so it would not seem as if he was in pain.

X and Y make love in her hotel room, the shutter open, a breeze on her left thigh. Someone is jangling keys in the corridor outside, and upstairs someone is singing a pop song. X says, I have to go now. His head is resting on a pillow inscribed with the name of the hotel; blue thread sewn into white cotton.

Is the settler X privileged and the wanderer Y deprived? For X and his partner Z, settled in the country of each other, there is something called a future.

For the deprived there is no word called future.
For the privileged there is only the future.

For the deprived the present is full of the absence of privilege.

For the privileged the present is full of the absence of deprivation.

Or is the wanderer Y privileged? Both intimate and alien, she can touch the world with a phantom hand.

X returns home to a chicken cooked by Z. Y eats take-away pizza on her balcony and washes her hands with a tiny square of perfumed soap inscribed with the hotel's name.

    She packs her bag.

Each new journey is a mourning for what has been left behind. The wanderer sometimes tries to recreate what has been left behind, in a new place. This always fails. To muster courage and endurance for a journey, it has to be embarked on with something like ambivalence. To retreat is to wane, fade, shrink, get less. This suggests that the privileged, who are not used to retreating, swell, expand and get more.

X rings through to the hotel and asks Y to stay one more night.

She, the wanderer, bum, refugee and rambler, drinks scotch on the rocks in a long glass, sitting in the corner of a bar. She smokes a cigarette bought in a small kiosk whilst changing trains at the last border she crossed. A border is an undefined margin between two things, sanity and insanity, for example. It is an edge. To be marginal is to be not fully defined. This thought excites her.

Although she is drinking whisky and smoking tobacco in a crowded bar in a strange city, and although she spends most of her time in cities, she often wants to be near water, to be under the stars, to feel the wind on her cheeks and wrists, to collect cones and kernels, shells, fossils, pebbles. She returns to the city with a bag full of these things. When she empties them out onto a carpet or floorboards or the wooden surface of a table, she puts her hand over them and they pulse into her palm. It is then she wonders if these are the things that give her health and special endurance. Some become mementos, some gifts, and the most special become talismans.

So now in the palm of her hand is a small brown feather.

X walks toward her. They drink more whisky. He tells her how glad he is she's stayed on an extra night. He talks and she listens. He says, 'I feel very sad tonight.' In fact she too is feeling sad, but he has not considered that this might be so and tells her a tale. While she is hearing why he is sad, she is also hearing voices from the radio blaring behind the bar. There has been an earthquake somewhere. A man is weeping. He is saying through an interpreter, 'I lost my wife's mother, my four children and my house. They are all gone.' The interpreter describes objects buried in the rubble. X does not hear these voices and he does not hear the unspoken sad voice of Y.

'Let's go for a swim.'

'It's dark,' he says.

'I know.'

He, the settler, present, visible and somewhere, is reluctant to swim with the wanderer at a strange hour. He has a home and he has Z to return to. He will return wet and Z will ask him why. How is he to relate the small intimate moments of his day to his co-settler? 'I drank whisky in a bar with my lover, and then we went for a swim.'

After a while he says, 'I'll watch you.'

While she swims and he watches her, she, like the privileged, will be perceptible, observed, present to the eye, witnessed. What she does not know is that he will watch her swim, and he will make her up. To him at that moment her eyelashes are blue, her wrists jangle with silver bands of alchemy and her hair is oiled with jasmine. In this way he will have stripped her of the possibility of being sad, needing things, having a past, present and future.

The sound of men and women wailing, and interpreters murmuring above them, leaks out of the radio.

They leave the bar. Sweepers in orange jackets clean the gutters. He puts his arm through hers and she strokes it with the small brown feather she has been holding all night. They walk out of town, across a road and then under a small bridge. Swans sleep on the river, necks tucked under their wings, floating silently under the moon.

She takes off her clothes. He watches her some of the time and he watches the swans. She is standing on the muddy bank. She is naked. He is clothed. X the settler has chosen not to swim. Y the wanderer is going to dive into the water and she is going to swim.

'Take off your earrings.'

She gives them to him.

'Don't swallow the water,' he shouts.

She swallows and swallows the water. And as she swallows she swallows the possibility she will always be alone. Swallows the river that will flow into the sea that is made from other waters that have flowed from mountains and hills, that will leak into oceans. She swallows geography, learns to swim in changing tides and temperatures, learns different strokes of the arms and legs, learns to speak in many tongues. She does this because she has no choice but to do so, and she comes out of the river to find him there, holding her earrings in his hand, and he says, 'But they don't fit. Who are you?'

'Who are you?' he said, backing away from the creature that emerged, streaked in mud, speaking a language he did not understand. 'Lie with me a while,' she murmured, and he, entranced by her golden fur drying under the moon, understood the request and consented. They lay together on the river bank while the swans, curved into themselves, slept on, both thinking how odd the fragrance of the other was, the texture of their skin and hair, the way their lips and tongues met, the way their bodies joined together. Both kept trying to find something recognisable, familiar in the other, but they could not. This was frightening but it was also arousing, and they experimented shyly with ways of pleasing each other. All the while he held a pair of earrings in his hand and would not let go. 'Take me to meet your family,' she said, lying on top of him, her golden belly rising and falling. At that moment he lost his desire and tried to separate himself from her, and at that moment the swans swam to the bank of the river to watch them.

'Why do you hate me?' she said, staring into his very blue eye.

'I don't.'

'You do.'

'I want you to be someone else.'

After a pause, he said again, 'Who are you?'

'I am the stranger who desires you and I have come to convince you it is truly you, in all your particularity, whom I desire.'

This made him sob, great gulping tears that made him ugly, mouth open, face crumpled until he found that he was howling and had to press his face into the mud to stifle his cries.

She said again, 'It is truly you, in all your particularity, whom I desire,' and he looked up into the black snout of her nose, at her golden paws and pink eyes, and vomited, spewed out the contents of his stomach, mostly chicken, into a little pool by his feet.

X wakes up from this short dream to see Y come out of the water, naked, and he says, 'Hi. I've been waiting for you. Here are your earrings.' He gives them to her. The wind blows and her breath tastes of whisky and mud.

As she dries herself Y asks X what he did while she was swimming.

'I watched you swim and then I fell asleep,' he says. And then he tells her he must go because Z will be waiting up for him.

'Stay,' she says.

He is silent.

She knows that by asking him to stay, she has invited him to tell her about his politics, ceremonies, poetry and food. She has broken a rule. She knows that though she has wandered through the country of his person, Z has permanent residence there. She knows this because she, the wanderer, has to know her heart and she has to know his heart. She has broken a rule and she runs the risk of being deported.

'Why do you hate me?' she asks.

'I don't.'

'You do.'

After a while he says, 'I want you to be someone else.'

'Who do you want me to be?'

An hour later, X will sit in front of a log fire with Z.

'Tell me about your day,' he says.

Z runs her fingers through her short cropped hair and stares into the fire. With her other hand she strokes the cat on her lap. 'Today we got a postcard from B. Look.' She shows him the card, an image of a date palm on the edge of a desert, and in the distance, white thorns.

'It is as if the desert has been posted through our front door,' she says. 'All day I have been tasting and smelling the desert, imagining its light and scale. And when I did my shopping, I imagined B there, sitting on ripples of sand.' She looks at X, her co-settler. 'What did you do today?'

'I went to the office early and left late. Then I came home and ate chicken with you. Then I had a whisky in a bar and went for a walk by the river.' Her hand moves slowly across the cat on her lap. The half moons of her fingernails glisten in its fur. 'Come here,' he says and pulls her towards him. He kisses her eyelids and tickles her neck with a small brown feather.

'You're wearing your hair down,' she says, winding a silver strand round and round her finger. 'I've never seen it loose before. You look like a stranger.' And then she says, 'Why do you hate me?'

'I don't.'

'You do.'

After a while he says, 'I want you to be someone else.'

Z takes the feather from his hand and strokes her wrist. She is feeling homesick, here in her own country, in her own home, in the bosom of her family, with her co-settler X.

Y is wringing water from her hair.

'Who do you want me to be?' she asks.

He shrugs and unwinds the green ribbon from his pony tail. His hair falls down past his shoulders to his waist. He plaits her wet hair and ties it with the ribbon.

'I'm missing someone I've never met.' He smiles. And then he looks away.

Y puts on her thick wool jacket.

'Who are you anyway?' he asks, watching her put heavy gold hoops into her ears.

'I am a country disguised and made up, offering itself to tourists.'

She gives him her small brown feather.

'Here. A souvenir to take home with you.'

She is the wanderer, bum, émigré, refugee, deportee, rambler. But most of all she is the strolling player.

XYZ.

To name someone is to give them a country.
To give them a country is to give them an address.
To give them an address is to give them a home.
To give them a home is to give them a wardrobe.
To give them a wardrobe is to give them a mirror.

What does Y see in her mirror?

Her hair is wet and it is tied with a green ribbon. She wears golden hoops in her ears. If she is a character, is she dressed for the part? What part is she going to play? Or perhaps she is dressed out of character? Dressed in a way one would not expect her to be. What is going to happen to Y?

The wardrobe with the mirror inside it belongs to the EUROPA HOTEL. Y can see she is shivering in a woollen jacket. If she is a persona, has she adopted a system, constructed a voice to speak through? Is Y a first, second or third person? A first person does the speaking. A second person is spoken to. A third person is spoken about.

What are Y's voices? The telephone rings and the receptionist puts through the call.

At this moment Z is re-arranging the furniture in her front room. Her co-settler X is at work. He is in a public place and she is in a private place. So now she is re-arranging the unspoken patterns of their privacy. First she moves the sofa away from the wall and places it by the window. Now she picks up objects that have fallen under the sofa over the years. Coins, a cigarette lighter, a biro, receipts, buttons, an elastic band, two magazines, bus tickets, keys, shopping lists, and lastly, a piece of paper folded into a neat square. She unfolds it and sees her co-settler's handwriting. He has written down the name of a hotel, the name of a bar, and a time. Z puts this into her pocket. Now she moves the rug from the centre of the room and places it at a diagonal somewhere else. She moves a table, chairs, three pictures, objects that have become so familiar she and her co-settler X have ceased to notice them, and places them somewhere unfamiliar. All the time she finds small things, long-lost and forgotten, out-of-date postage stamps, X's cuff-links, an old lipstick. They have lived a life together and Z collects it like evidence. Her short-cropped black hair is covered in dust. The cat, upset by the re-arranging, runs across the room and back again, looking for places to hide. Z calls out to it. And then B walks in.

X is about to go into a meeting. He is collecting the paperwork he needs with one hand, and speaking on the telephone with the other. A colleague mimes something and points to his watch. X nods. A secretary brings him a cup of coffee. He mimes thank you. Through the glass windows of the open-plan office he can see bodies at work. Sitting, standing, leaning. Another colleague hands him a wad of papers which he flicks through whilst speaking on the telephone, considers this new information, stands up and reaches for his jacket, all the time still speaking on the telephone. Someone knocks on the glass window and tells him to hurry up. This is also mimed.

Mim-e-sis: Imitation. Representation. An ancient farcical play of real life with mimicry. A person's supposed or imagined words.

The person X is speaking to is Y.

Y is in her hotel room, bags packed, speaking on the telephone to X.
    X: I've called to say goodbye.
    Y: Goodbye.

X: I can't speak here. I'm about to go to a meeting.

Y: Bye then.

X: Don't go without saying goodbye properly.

Y: What does saying goodbye properly mean?

X: Can I call you later?

Y: I won't be here.

X: Where will you be? Give me a telephone number.

Y: Why haven't you asked me that before?

X: Look, I have to go.

Y: Bye then.

Z is carrying an armful of books when she sees B.

'You're moving things,' B says. 'Why?'

Z drops the books on the newly positioned sofa and walks over to her. 'I got a postcard from you yesterday. I can't believe you're here. I thought you were there. I've been imagining you there.'

B is tanned. Her shoulders are smooth and brown. 'I got back three days ago.' She looks around at the front room. 'I've been imagining this room,' she says. 'I knew exactly where the table is, or was. You sat there,' she points, 'and I sat here. But you've changed it.' She smiles. 'It's like waking up on the wrong side of the bed.' She runs her fingers through Z's short hair. 'Oh honey,' she says. They kiss in the disordered room. This is the first time they have kissed. Like that. Slow and long. Z strokes B's bare shoulder and then moves her hand to her breasts.

X is having a drink after work with a male colleague. 'I

think I am in love,' he says. 'But I don't want to hurt my wife.'

Y is about to get on a train. She is returning to the city she knows better than other cities. She knows how its transport works, where its most obscure coffee bars are hidden, opening and closing times, its swimming pools, banks, night-clubs, cheap restaurants, theatres and cinemas.

'Oh honey.'
  'You're lovely.'
  'Oh.'
  'There.'
  'That's nice.'
  'Like that?'
  'Hmm.'
  'Oh.'
  'Sweetheart.'
  'Hmm.'
  'There.'
  'Is that nice?'
  'Harder.'
  'Oh.'
  'Oh baby.'
  'Oh.'
  'Aren't you beautiful?'

X walks in to find his co-settler Z naked in the arms of their co-friend B who at this moment has her head buried between the thighs of Z. In addition this is happening in his front room which has been re-arranged and he does not know where anything is. Nor, it occurs to him does he know who his co-settler is, how she might want to be touched, or where she wants to live, because he hears these words fall from her lips into the crease of B's brown neck.

'Take me somewhere else.'

Y is lying in a bath in the place she calls her home in the city she knows best. She is listening to music and the kettle is boiling. On the small desk next to her computer, she has placed five white goose feathers in a tin can. Her clothes hang on a rail. Her bed-linen is familiar and cool. Her cupboard is full of spices in labelled glass jars. She knows every book on her shelf. She knows what the view out of her window will be. So now she is lying in her bath looking out of the window and she knows that at about nine o'clock she will meet a friend and they will eat out somewhere in the centre of the city. They will link arms and walk through the traffic as if they have nine lives.

Z looks up into the eyes of her co-settler X.

His hands are in his pockets.

B puts on her T-shirt and smiles at X.

'Hello,' she says. And then she zips up her jeans and asks Z if she has any wine in the house.

'No,' X says.

'Yes,' Z says. She slips her hands into the pocket of her silk blouse, bought with X's money, and gives him the piece of paper folded into a small square she found that morning. X unfolds it. It says HOTEL EUROPA, BAR LEONARDO, 6PM.

'Go and find her,' Z says.

B calls out to the cat who is hiding under the table.

After a while, X says, 'I don't know where she is.'

Y is eating pasta in a café in the centre of the city with her friend. It is busy and hot and the tables are full of pimps, lovers, prostitutes, friends, students, tourists and loners reading journals. They stare, talk, fight, shout out for ashtrays, ask for advice, borrow money, joke, cry, tell lies, describe their families, romances, children, and some even tell their companion that they love them.

What does Y possess? She who owns no property, has no inheritance, husband, children, savings or pension plan? Y is a first, second and third person. These are all her voices. XYZ. Z and X will reconcile their differences, buy things for their home and grow old together. Will Y grow old? Have we met her elderly and survived and occupying public space, not with melancholy or eccentricity, but as a matter of fact?

Y dips her bread in her wine.

'Fear is an invented thing,' she says.

'How do you uninvent it then?' J.K. asks in Spanish.

'You ask too many questions,' Y replies in English.

At that moment B walks in and they invite her to sit at their table.